STEMD^2 Research & Development Group
Center on Disability Studies
College of Education
University of Hawaiʻi at Mānoa

http://stemd2.com/

 First published (2019) by the Center on Disability Studies, University of Hawaiʻi at Mānoa, Honolulu, Hawaiʻi. This document is based upon work supported by the Department of Education, Native Hawaiian Education Act Program under award #S362A180011. Any opinions, findings, and conclusions or recommendations expressed in this material are those of the authors and do not necessarily reflect the views of the United States Department of Education. For further information about this document and Ne'epapa Ka Hana 2.0 project, please contact Dr. Kaveh Abhari at abhari@hawaii.edu.

ISBN: 978-0-9983142-6-6
First release, 2019

Ne'epapa Ka Hana Sixth-Grade Mathematics Resources

Let's Go from Mauka to Makai

Student Activities

Project Director	Kaveh Abhari
Content Developers	Robert G. Young Justin S. Toyofuku
Creative Designer	MyLan Tran
Publication Designer	Robert G. Young

Acknowledgments

We would like to thank Kelli Ching, Crystal Yoo, Katy Parsons, and Robyn Rice for advising on middle school mathematics. Thank you Moa Viebke, Nohea Behler, and Robyn Rice for significant help reviewing and editing. Mahalo nui to Moa Viebke and Nohea Behler for major contributions in writing the introductions.

Contents

Introduction to *Let's Go from Mauka to Makai*

Let's Go from Mauka to Makai has been created by Ne'epapa Ka Hana (NKH) for 6th grade students of Hawai'i. The mathematical activities featured in *Let's Go from Mauka to Makai* have been constructed to enrich students mathematical abilities through culturally responsive material and to increase interest and participation in the mathematical classroom.

Through the collaboration between the Ne'epapa Ka Hana (NKH) 2.0 and STEMD2, the mathematical curriculum of *Let's Go from Mauka to Makai* has been made available for online classroom integration. The math activities are offered on a social platform which allows for online collaboration between students and teachers. The platform is free and accessible to users with internet access at www.community.stemd2.com

Ne'epapa Ka Hana (NKH) develop programs and materials that are culturally responsive to Hawai'i's unique diversity – using research-based practices to support ongoing STEM education efforts across the state. *Let's Go from Mauka to Makai* is the 6th addition to NKH's middle-school book series.

The *Let's Go from Mauka to Makai* book is composed by mathematical learning activities that serve as a supplementary mathematics curriculum to enhance inclusive instruction based on problem-based learning strategies and connectivism principles. The math activities incorporate real-world challenges that reflect Hawai'i's unique culture, society, and geography. The goal of these activities is for both kumu and haumāna to collaborate while thinking critically and creatively – thereby, deepening students' understanding, application, and appreciation for mathematical thinking in the context of Hawaiian and island culture.

Let's Go from Mauka to Makai, highlights the Common Core State Standards through the theme of Hawaiian environment and ecology. The curriculum focuses on supporting Hawaiian's natural habitat, evaluating current environmental threats and illustrate how to incorporate mathematical solutions to solve them.

Common Core State Standards

Common Core State Standard	***Let's Go from Mauka to Makai* Unit**
Ratios & Proportional Relationships	
6.RP.1 Understand the concept of a ratio and use ratio language to describe a ratio relationship between two quantities. *For example, "The ratio of wings to beaks in the bird house at the zoo was 2:1, because for every 2 wings there was 1 beak." "For every vote candidate A received, candidate C received nearly three votes."*	3
6.RP.2 Understand the concept of a unit rate a/b associated with a ratio a:b with b $\neq$ 0, and use rate language in the context of a ratio relationship. *For example, "This recipe has a ratio of 3 cups of flour to 4 cups of sugar, so there is 3/4 cup of flour for each cup of sugar." "We paid $75 for 15 hamburgers, which is a rate of $5 per hamburger."* (Note: Expectations for unit rates in this grade are limited to non-complex fractions.)	3
6.RP.3 Use ratio and rate reasoning to solve real-world and mathematical problems, e.g., by reasoning about tables of equivalent ratios, tape diagrams, double number line diagrams, or equations.	3
6.RP.3a Make tables of equivalent ratios relating quantities with whole-number measurements, find missing values in the tables, and plot the pairs of values on the coordinate plane. Use tables to compare ratios.	3
6.RP.3b Solve unit rate problems including those involving unit pricing and constant speed. *For example, if it took 7 hours to mow 4 lawns, then at that rate, how many lawns could be mowed in 35 hours? At what rate were lawns being mowed?*	3
6.RP.3c Find a percent of a quantity as a rate per 100 (e.g., 30% of a quantity means 30/100 times the quantity); solve problems involving finding the whole, given a part and the percent.	3

Continued on next page

Common Core State Standard	*Let's Go from Mauka to Makai* Unit
6.RP.3d Use ratio reasoning to convert measurement units; manipulate and transform units appropriately when multiplying or dividing quantities.	3
The Number System	
6.NS.1 Interpret and compute quotients of fractions, and solve word problems involving division of fractions by fractions, e.g., by using visual fraction models and equations to represent the problem. *For example, create a story context for (2/3) ÷ (3/4) and use a visual fraction model to show the quotient; use the relationship between multiplication and division to explain that (2/3) ÷ (3/4) = 8/9 because 3/4 of 8/9 is 2/3. (In general, (a/b) ÷ (c/d) = ad/bc.). How much chocolate will each person get if 3 people share 1/2 lb of chocolate equally? How many 3/4-cup servings are in 2/3 of a cup of yogurt? How wide is a rectangular strip of land with length 3/4 mi and area 1/2 square mi?*	2
6.NS.2 Fluently divide multi-digit numbers using the standard algorithm.	2
6.NS.3 Fluently add, subtract, multiply, and divide multi-digit decimals using the standard algorithm for each operation.	2
6.NS.4 Find the greatest common factor of two whole numbers less than or equal to 100 and the least common multiple of two whole numbers less than or equal to 12. Use the distributive property to express a sum of two whole numbers 1-100 with a common factor as a multiple of a sum of two whole numbers with no common factor. *For example, express 36 + 8 as 4 (9 + 2).*	1
6.NS.5 Understand that positive and negative numbers are used together to describe quantities having opposite directions or values (e.g., temperature above/below zero, elevation above/below sea level, debits/credits, positive/negative electric charge); use positive and negative numbers to represent quantities in real-world contexts, explaining the meaning of 0 in each situation.	1
6.NS.6 Understand a rational number as a point on the number line. Extend number line diagrams and coordinate axes familiar from previous grades to represent points on the line and in the plane with negative number coordinates.	1, 5
6.NS.6a Recognize opposite signs of numbers as indicating locations on opposite sides of 0 on the number line; recognize that the opposite of the opposite of a number is the number itself, e.g., −(−3) = 3, and that 0 is its own opposite.	1

Continued on next page

Common Core State Standard	*Let's Go from Mauka to Makai* Unit
6.NS.6b Understand signs of numbers in ordered pairs as indicating locations in quadrants of the coordinate plane; recognize that when two ordered pairs differ only by signs, the locations of the points are related by reflections across one or both axes.	5, 6
6.NS.6c Find and position integers and other rational numbers on a horizontal or vertical number line diagram; find and position pairs of integers and other rational numbers on a coordinate plane.	1, 5
6.NS.7 Understand ordering and absolute value of rational numbers.	1
6.NS.7a Interpret statements of inequality as statements about the relative position of two numbers on a number line diagram. *For example, interpret −3 > −7 as a statement that −3 is located to the right of −7 on a number line oriented from left to right.*	1
6.NS.7b Write, interpret, and explain statements of order for rational numbers in real-world contexts. *For example, write −3° C > −7° C to express the fact that −3° C is warmer than −7° C.*	1
6.NS.7c Understand the absolute value of a rational number as its distance from 0 on the number line; interpret absolute value as magnitude for a positive or negative quantity in a real-world situation. *For example, for an account balance of −30 dollars, write |−30| = 30 to describe the size of the debt in dollars.*	1
6.NS.7d Distinguish comparisons of absolute value from statements about order. *For example, recognize that an account balance less than −30 dollars represents a debt greater than $30.*	1
6.NS.8 Solve real-world and mathematical problems by graphing points in all four quadrants of the coordinate plane. Include use of coordinates and absolute value to find distances between points with the same first coordinate or the same second coordinate.	5, 6
Equations & Expressions	
6.EE.1 Write and evaluate numerical expressions involving whole-number exponents.	4
6.EE.2 Write, read, and evaluate expressions in which letters stand for numbers.	4

Continued on next page

Common Core State Standard	***Let's Go from Mauka to Makai* Unit**
6.EE.2a Write expressions that record operations with numbers and with letters standing for numbers. *For example, express the calculation "Subtract y from 5" as 5 – y.*	4
6.EE.2b Identify parts of an expression using mathematical terms (sum, term, product, factor, quotient, coefficient); view one or more parts of an expression as a single entity. *For example, describe the expression 2 (8 + 7) as a product of two factors; view (8 + 7) as both a single entity and a sum of two terms.*	4
6.EE.2c Evaluate expressions at specific values of their variables. Include expressions that arise from formulas used in real-world problems. Perform arithmetic operations, including those involving whole-number exponents, in the conventional order when there are no parentheses to specify a particular order (Order of Operations). *For example, use the formulas* $V = s^3$ *and* $A = 6s^2$ *to find the volume and surface area of a cube with sides of length s = 1/2.*	4
6.EE.3 Apply the properties of operations as strategies to generate equivalent expressions. *For example, apply the distributive property to the expression 3(2 + x) to produce the equivalent expression 6 + 3x; apply properties of operations to y + y + y to produce the equivalent expression 3y.*	4
6.EE.4 Identify when two expressions are equivalent (i.e., when the two expressions name the same number regardless of which value is substituted into them). *For example, the expressions y + y + y and 3y are equivalent because they name the same number regardless of which number y stands for.*	4
6.EE.5 Understand solving an equation or inequality as a process of answering a question: which values from a specified set, if any, make the equation or inequality true? Use substitution to determine whether a given number in a specified set makes an equation or inequality true.	5
6.EE.6 Use variables to represent numbers and write expressions when solving a real-world or mathematical problem; understand that a variable can represent an unknown number, or, depending on the purpose at hand, any number in a specified set.	4, 5
6.EE.7 Solve real-world and mathematical problems by writing and solving equations of the form x + p = q and px = q for cases in which p, q and x are all nonnegative rational numbers.	5, 6

Continued on next page

Common Core State Standard	***Let's Go from Mauka to Makai* Unit**
6.EE.8 Write an inequality of the form $x > c$ or $x < c$ to represent a constraint or condition in a real-world or mathematical problem. Recognize that inequalities of the form $x > c$ or $x < c$ have infinitely many solutions; represent solutions of such inequalities on number line diagrams.	5
6.EE.9 Use variables to represent two quantities in a real-world problem that change in relationship to one another; write an equation to express one quantity, thought of as the dependent variable, in terms of the other quantity, thought of as the independent variable. Analyze the relationship between the dependent and independent variables using graphs and tables, and relate these to the equation. *For example, in a problem involving motion at constant speed, list and graph ordered pairs of distances and times, and write the equation $d = 65t$ to represent the relationship between distance and time.*	5
Geometry	
6.G.1 Find area of right triangles, other triangles, special quadrilaterals, and polygons by composing into rectangles or decomposing into triangles and other shapes; apply these techniques in the context of solving real-world and mathematical problems.	5, 6
6.G.2 Find the volume of a right rectangular prism with fractional edge lengths by packing it with unit cubes of the appropriate unit fraction edge lengths, and show that the volume is the same as would be found by multiplying the edge lengths of the prism. Apply the formulas $V = l\ w\ h$ and $V = b\ h$ to find volumes of right rectangular prisms with fractional edge lengths in the context of solving real-world and mathematical problems.	6
6.G.3 Draw polygons in the coordinate plane given coordinates for the vertices; use coordinates to find the length of a side joining points with the same first coordinate or the same second coordinate. Apply these techniques in the context of solving real-world and mathematical problems.	6
6.G.4 Represent three-dimensional figures using nets made up of rectangles and triangles, and use the nets to find the surface area of these figures. Apply these techniques in the context of solving real-world and mathematical problems.	6

Continued on next page

Common Core State Standard	***Let's Go from Mauka to Makai* Unit**
Statistics & Probability	
6.SP.1 Recognize a statistical question as one that anticipates variability in the data related to the question and accounts for it in the answers. *For example, "How old am I?" is not a statistical question, but "How old are the students in my school?" is a statistical question because one anticipates variability in students' ages.*	7
6.SP.2 Understand that a set of data collected to answer a statistical question has a distribution which can be described by its center, spread, and overall shape.	7
6.SP.3 Recognize that a measure of center for a numerical data set summarizes all of its values using a single number, while a measure of variation describes how its values vary using a single number.	7
6.SP.4 Display numerical data in plots on a number line, including dot plots, histograms, and box plots.	7
6.SP.5 Summarize numerical data sets in relation to their context, such as by:	7
6.SP.5a Reporting the number of observations.	7
6.SP.5b Describing the nature of the attribute under investigation, including how it was measured and its units of measurement.	7
6.SP.5c Giving quantitative measures of center (median and/or mean) and variability (interquartile range and/or mean absolute deviation), as well as describing any overall pattern and any striking deviations from the overall pattern with reference to the context in which the data were gathered.	7
6.SP.5d Relating the choice of measures of center and variability to the shape of the data distribution and the context in which the data were gathered.	7

Unit 1: Numbers

In this unit, we'll learn how to use positive and negative integers, fractions, and decimals to describe situations through Hawaiʻi's vast biodiversity and unique ecosystem. There are four activities in this unit. *Module 1* involves working with integers to find species at different altitudes on the Islands of Hawaiʻi, *Module 2* focuses on how to use factors and multiples to support our native environment, and *Module 3* explores the rain seasons of Hawaiʻi with the help of rational numbers. The final activity is cumulative and incorporates concepts from each of the previous activities in this unit.

Common Core State Standards

Common Core State Standard	Module 1	Module 2	Module 3	Unit 1
6.NS.4 Find the greatest common factor of two whole numbers less than or equal to 100 and the least common multiple of two whole numbers less than or equal to 12. Use the distributive property to express a sum of two whole numbers 1-100 with a common factor as a multiple of a sum of two whole numbers with no common factor. *For example, express 36 + 8 as 4 (9 + 2).*		X		X
6.NS.5 Understand that positive and negative numbers are used together to describe quantities having opposite directions or values (e.g., temperature above/below zero, elevation above/below sea level, debits/credits, positive/negative electric charge); use positive and negative numbers to represent quantities in real-world contexts, explaining the meaning of 0 in each situation.	X		X	X
6.NS.6 Understand a rational number as a point on the number line. Extend number line diagrams and coordinate axes familiar from previous grades to represent points on the line and in the plane with negative number coordinates.	X		X	X
6.NS.6a Recognize opposite signs of numbers as indicating locations on opposite sides of 0 on the number line; recognize that the opposite of the opposite of a number is the number itself, e.g., −(−3) = 3, and that 0 is its own opposite.	X		X	X
6.NS.6c Find and position integers and other rational numbers on a horizontal or vertical number line diagram; find and position pairs of integers and other rational numbers on a coordinate plane.	X		X	X
6.NS.7 Understand ordering and absolute value of rational numbers.	X		X	
6.NS.7a Interpret statements of inequality as statements about the relative position of two numbers on a number line diagram. *For example, interpret −3 > −7 as a statement that −3 is located to the right of −7 on a number line oriented from left to right.*	X		X	X

Continued on next page

Common Core State Standard	Module 1	Module 2	Module 3	Unit 1
6.NS.7b Write, interpret, and explain statements of order for rational numbers in real-world contexts. *For example, write −3° C > −7° C to express the fact that −3° C is warmer than −7° C.*	X		X	X
6.NS.7c Understand the absolute value of a rational number as its distance from 0 on the number line; interpret absolute value as magnitude for a positive or negative quantity in a real-world situation. *For example, for an account balance of −30 dollars, write |−30| = 30 to describe the size of the debt in dollars.*	X		X	
6.NS.7d Distinguish comparisons of absolute value from statements about order. *For example, recognize that an account balance less than −30 dollars represents a debt greater than $30.*	X		X	X

Module 1: Integers *Activity*

Biodiversity refers to how many different kinds of living organisms are present in an area. For example, a farm with 1000 chickens has less biodiversity than a farm with a chicken, a duck, a cow, and a pig. The Hawaiian Islands are known for having incredible biodiversity. From the bottom of the ocean to the top of the mountains, we can find a large variety of plants, animals, and other types of life.

Let's take a few days to travel the Big Island of Hawai'i. We will write down some of the life we see and the altitude (height) above sea level that we see it.

We will start the week by exploring the mountains. On the top of Mauna Kea we find a rare Wēkiu beetle at an altitude of 3725 meters. At an altitude of 1432 meters on a nearby mountain, we see the state bird of Hawai'i, the Nēnē goose. On the way down, we see an 'ōhi'a lehua plant with beautiful flowers at an altitude of 1202 meters.

Wēkiu beetle

Nēnē goose

'Ōhi'a lehua

On Friday, we sail out into the ocean. You lower a fishing line down to -642 meters and catch an opakapaka. Your friend lowers a line as deep as she can. Her line goes down 1432 meters below sea level, and she pulls up a skinny fish, with a huge mouth, called a gulper eel.

Opakapaka

Gulper eel

The following weekend, we visit a white sandy beach. While swimming offshore, we see a tiny krill at an altitude of about -1 meter (or 1 meter below sea level). At sea level, we spot a Portuguese man o' war floating around. We swim around it and carefully climb 1 meter up a few slippery rocks to collect some delicious opihi.

Krill

Portuguese man o' war

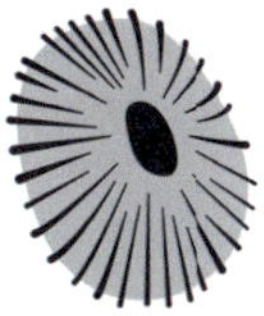
Opihi

1. Use the following table to write down the names of the plants and animals that we saw and the altitude that we saw them. Be sure to use negative numbers for animals that were found below sea level.

Plant/animal name	Altitude spotted (meters)

2. Write down the name of the plants or animals in a list from **lowest altitude to highest altitude**.

(Lowest altitude) (Highest altitude)

3. Write down the name of the plants or animals from **closest to furthest from sea level** (0 meters). If two plants/animals have the same distance from sea level, then write them next to each other.

(Closest to sea level) (Furthest from sea level)

4. With some practice, many people are able to dive down to **10** meters below sea level or hike up to an altitude of 1500 meters without much effort. Let x be the altitudes that we can go to.

   (a) If we cannot go any lower than 10 meters below sea level, then what are the altitudes that we **can** go to?
   (Circle one): $x \geq 10$ $x \leq 10$ $x \geq -10$ $x \leq -10$

   (b) If we cannot go any higher than an altitude of 1500 meters, then what are the altitudes that we **can** go to?
   (Circle one): $x \geq 1500$ $x \leq 1500$ $x \geq -1500$ $x \leq -1500$

   (c) List the animals that could have been spotted just by diving or hiking.

5. Which of these plants and animals have you seen in real life? Where did you see them?

6. What is the most interesting plant or animal that you have seen in Hawai'i? What was so interesting about it?

Module 2: Factors and Multiples *Activity*

To improve the condition of our environment, the government has decided to protect a large slice of land called an ahupua'a. All ahupua'a are made up of two parts, the mauka and the makai. The mauka part is close to the mountains and the makai part is close to the sea. We don't know the exact measurements of the protected ahupua'a, but we do know that the **mauka part has an area of 128 square kilometers**, and the **makai part has an area of 96 square kilometers**. We also know that the ahupua'a and its parts are shaped like rectangles.

For example, if the protected ahupua'a is 8 kilometers wide, then the mauka part must be 16 kilometers long and the makai part must be 12 kilometers long.

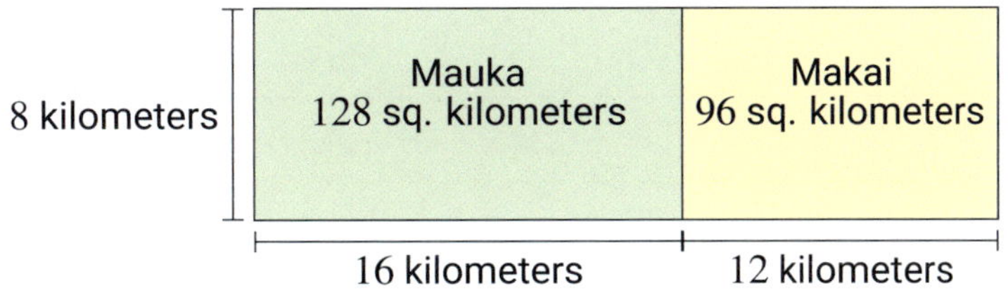

For another example, if the protected ahupua'a is only 4 kilometers wide, then the mauka part must be 32 kilometers long and the makai part must be 24 kilometers long.

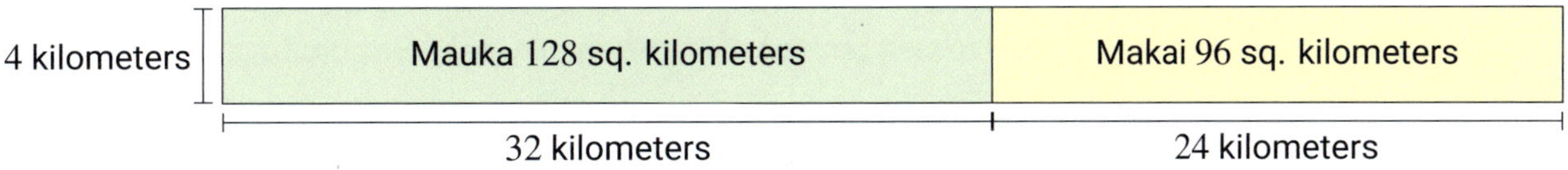

1. Suppose that the width is a **whole number**, draw at least two more possible rectangles to represent the protected area. Be sure to label the sides of your drawings just like in the example above.

2. If the width is a **whole number**, what is the largest width that the rectangle can be?

The government is asking for your help in taking care of this area by visiting it regularly. They want you to visit the mauka part **once every 8 weeks** and let them know if that area is doing well. They also want you the visit the makai part **once every 3 weeks** and let them know if it is doing well. This week, you will check on both the mauka and makai areas.

3. Sometimes you will have to visit both parts in the same week. **How often** does this happen?

4. If you needed to take care of the ahupua'a for the next 80 weeks, during which weeks would you visit both parts? **How many times** will you need to visit both parts in the same week? Be sure to count this week as the first time.

5. If you visited these protected areas, what would you look for to determine if the area is doing okay? Why would you look for these specific things?

Module 3: Rational Numbers *Activity*

Mount Wai'ale'ale on Kaua'i is one of the rainiest places on Earth. Normally, rainfall is measured in inches, but since it rains so much on Mount Wai'ale'ale, we can measure it in feet.

Mount Wai'ale'ale

We want to compare the rainfall on Mt. Wai'ale'ale with another rainy place called Mawsynram ("mah-sin-rem") in India by calculating the difference. Let's take the rainfall from Mt. Wai'ale'ale and subtract the rainfall from Mawsynram. Here is the data.

	Rainfall (feet)		
Season	Mt. Wai'ale'ale	Mawsynram	Difference
Spring	$8\frac{1}{2}$	$6\frac{1}{5}$	$2\frac{3}{10}$
Summer	7.3	21.5	−14.2
Fall	$7\frac{3}{5}$	$4\frac{2}{5}$	$3\frac{1}{5}$
Winter	7.05	0.2	6.85

1. In the table above, are there any pairs of numbers that are exactly opposite? If so, which pairs?

2. For each season, plot the "difference" column on the number line. (Do not plot any of the other numbers on the table.)

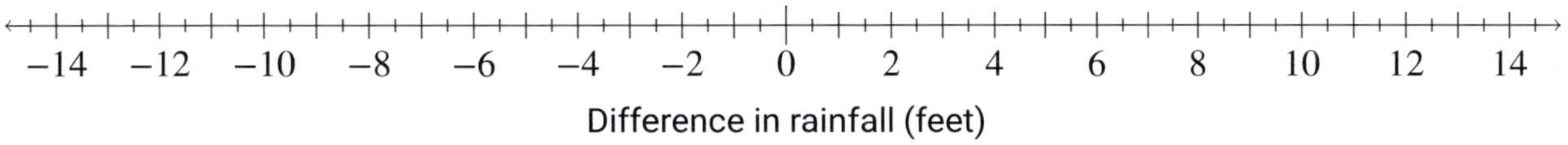

3. When the difference in rain is 0, both Mt. Wai'ale'ale and Mawsynram have the same amount of rainfall. What about when the difference is negative or positive?

Mt. Wai'ale'ale / Mawsynram has more rainfall when the difference is **negative**.
(circle one)

Mt. Wai'ale'ale / Mawsynram has more rainfall when the difference is **positive**.
(circle one)

4. Which season has the largest difference between rainfall in Mount Wai'ale'ale and Mawsynram? How can you tell?

5. Which season has the smallest difference between rainfall in Mount Wai'ale'ale and Mawsynram? How can you tell?

6. Let's compare the "difference" in the summer with the seasons that are not summer.

   (a) Write down the difference in rainfall in the summer.

   (b) Add up all of the differences from **spring**, **fall**, and **winter** to find the difference of the seasons that are not summer.

7. In one year, does it rain more in Mount Wai'ale'ale or Mawsynram? How can you use part 6 to help you answer this question?

8. Of these two places, which one would you consider to have "the most unusual" weather and why?

Unit 1: *Cumulative Activity*

Ancient Hawai'i used to be full of strong koa trees and fragrant 'iliahi (sandalwood) trees, but centuries of deforestation have made most of these trees and the animals that depend on them disappear. Now volunteers are working to bring the trees back.

Koa sapling

'Iliahi sapling

Kahā and Kekai are two gardeners who specialize in growing saplings (baby trees) of native plants. Kahā plans to donate 180 koa saplings, and Kekai will donate 120 'iliahi saplings. They will put them into packages and give them to volunteers who will plant and take care of them. Each package will contain the same mix of saplings, and no saplings will be left over once the packages are filled.

1. Kahā and Kekai are trying to decide how many koa and 'iliahi saplings to put in each package. For example, if each package has 18 koa and 12 'iliahi, then they can make 10 packages.

   Work with a partner to list **other possible ways** to package the koa and 'iliahi.

Koa per package **18** 'Iliahi per package **12** Number of packages **10**	Koa per package 'Iliahi per package Number of packages	Koa per package 'Iliahi per package Number of packages
Koa per package 'Iliahi per package Number of packages	Koa per package 'Iliahi per package Number of packages	Koa per package 'Iliahi per package Number of packages
Koa per package 'Iliahi per package Number of packages	Koa per package 'Iliahi per package Number of packages	Koa per package 'Iliahi per package Number of packages
Koa per package 'Iliahi per package Number of packages	Koa per package 'Iliahi per package Number of packages	Koa per package 'Iliahi per package Number of packages

2. What is the maximum number of packages Kahā and Kekai can make?

After receiving the donations, the volunteers begin to plant their saplings. Over the next few weeks, many trees will be planted but some will die. Below is a table that shows how the number of trees have changed since the saplings were planted. For example, "15" means that there are 15 more trees now than before the saplings were planted. "−5" means that there are 5 less trees now than before the saplings were planted.

Week	Change in number of trees
0	0
1	3
2	−2
3	−6
4	−5
5	1
6	6

3. For each week, plot the "change in number of trees" on the number line. Make sure to label the number line.

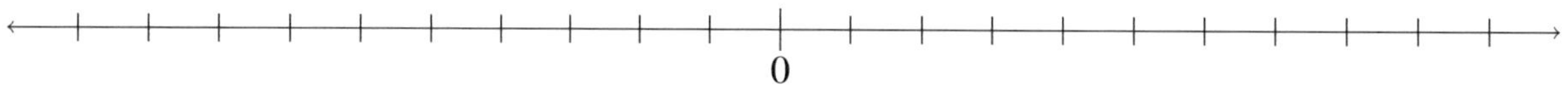

4. Determine whether the following statements are **true or false**.
   (a) There were more trees in week 3 than week 2. True or False
   (b) There were more trees in week 4 than week 3. True or False
   (c) There were more trees in week 3 than week 6. True or False
   (d) There were more trees in week 5 than week 4. True or False
   (e) Two of the numbers in "change in number of trees" column are opposite. True or False

5. Restoring a forest doesn't always work as planned. A new sapling has to fight for sun, water, and nutrients while facing bad weather, pollution, and pests. If you wanted to learn how to grow a plant, what kind of plant would you choose and why? Please share your thoughts with your partner or in the online comment section.

Unit 2: Number Operations

In this unit, we'll learn how to use positive integers, fractions, and decimals to solve problems through exploration of invasive species and evaluating Hawai'i's ecosystem. There are three activities in this unit. *Module 4* involves tracking and evaluating the coconut beetle through the use of operations with fractions. *Module 5* focuses on operations with decimals by evaluating the rain fall on a ahupua'a. The final activity is cumulative and incorporates concepts from each of the previous activities in this unit.

Common Core State Standards

Common Core State Standard	Module 4	Module 5	Unit 2
6.NS.1 Interpret and compute quotients of fractions, and solve word problems involving division of fractions by fractions, e.g., by using visual fraction models and equations to represent the problem. *For example, create a story context for (2/3) ÷ (3/4) and use a visual fraction model to show the quotient; use the relationship between multiplication and division to explain that (2/3) ÷ (3/4) = 8/9 because 3/4 of 8/9 is 2/3. (In general, (a/b) ÷ (c/d) = ad/bc.). How much chocolate will each person get if 3 people share 1/2 lb of chocolate equally? How many 3/4-cup servings are in 2/3 of a cup of yogurt? How wide is a rectangular strip of land with length 3/4 mi and area 1/2 square mi?*	X		X
6.NS.2 Fluently divide multi-digit numbers using the standard algorithm.		X	X
6.NS.3 Fluently add, subtract, multiply, and divide multi-digit decimals using the standard algorithm for each operation.		X	X
6.NS.4 Find the greatest common factor of two whole numbers less than or equal to 100 and the least common multiple of two whole numbers less than or equal to 12. Use the distributive property to express a sum of two whole numbers 1-100 with a common factor as a multiple of a sum of two whole numbers with no common factor. *For example, express 36 + 8 as 4 (9 + 2).*	X		X

Module 4: Operations with Fractions *Activity*

A $13\frac{1}{2}$ acre forest has been attacked by invasive coconut rhinoceros beetles. A team of 8 researchers have decided to spread out and check on the health of the forest.

Coconut rhinoceros beetle

1. Each of the 8 researchers have been assigned a equal share of the $13\frac{1}{2}$ acres of forest. This means that they have to watch $1\frac{11}{16}$ or $\frac{27}{16}$ acres of forest each. How did they come up with this number? Please show all of your work below.

Sometimes, we have different ways to describe the same thing. For example, when we say that "we lost half of our money," we can describe this by taking the amount of money we had and **dividing** by 2. We can also take the money we had and **multiply** by $\frac{1}{2}$. It's the same thing.

2. After watching the forests for several years, each researcher saw that their part has gotten smaller. They wrote down a **number** that describes how much the forest has changed, but they didn't write down whether you are supposed to **multiply or divide** by that number. Write down the missing sign (× or ÷) that makes the forest size ($\frac{27}{16}$) **smaller**. Find the size of the smaller forest part. Give your answer as a mixed number.

   (a) Researcher 1: $\frac{27}{16} \bigcirc\!\!\!\!\div\ \frac{4}{3} = 1\frac{17}{64}$

   $$\frac{27}{16} \div \frac{4}{3} = \frac{27}{16} \times \frac{3}{4} = \frac{81}{64} = 1\frac{17}{64}$$

   (b) Researcher 2: $\frac{27}{16} \bigcirc \frac{16}{25} =$

   (c) Researcher 3: $\frac{27}{16} \bigcirc \frac{8}{15} =$

(d) Researcher 4: $\frac{27}{16} \bigcirc \frac{27}{20} =$

(e) Researcher 5: $\frac{27}{16} \bigcirc \frac{18}{5} =$

(f) Researcher 6: $\frac{27}{16} \bigcirc \frac{25}{24} =$

(g) Researcher 7: $\frac{27}{16} \bigcirc \frac{16}{21} =$

(h) Researcher 8: $\frac{27}{16} \bigcirc \frac{21}{20} =$

Module 5: Operations with Decimals *Activity*

If you have taken a trip around a Hawaiian island, you may have noticed that the weather changes between different ahupua'a. Even neighboring ahupua'a can have very different amounts of wind, rain, and even sunlight. Let's look at three different ahupua'a on a day that has been raining on and off. We recorded the time of each rain shower and the total amount of rain that fell by the end of the day.

Ahupua'a 1

Rain times	1.55 hours 0.12 hours 1 hour
Rain amount	2.1 inches

Ahupua'a 2

Rain times	0.2 hours 5.15 hours 9.8 hours
Rain amount	4.0 inches

Ahupua'a 3

Rain times	0.8 hours 0.5 hours 0.2 hours
Rain amount	2.9 inches

1. On which ahupua'a did the most rain fall?

2. For each ahupua'a, calculate the total rain time. Then rank each ahupua'a from **least to greatest** total time spent under rain.

3. The **rain intensity** tells us whether it is raining lightly or whether it is raining heavily (known as "raining cats and dogs"). **Rain intensity** can be found by dividing the total amount of rain by the total rain time. Find the rain intensity for each ahupua'a. Round to two decimal places. Be sure to label your answer in inches per hour.

4. Discuss with a partner which ahupuaʻa has the lightest rain and which has the heaviest.
5. For each ahupuaʻa, draw a picture of it raining outside. Make sure it matches your discussion from part 4.

Ahupuaʻa 1

Ahupuaʻa 2

Ahupuaʻa 3

Unit 2: *Cumulative Activity*

The coconut rhinoceros beetle is an invasive beetle that is spreading and killing both coconut and palm trees. As of 2019, the rhinoceros beetle has been found in the ʻEwa moku of Oʻahu, mostly around Pearl Harbor. (A moku is a group of ahupuaʻa.)

Coconut rhinoceros beetle

You are monitoring a small palm forest on Oʻahu that is under attack by the rhinoceros beetle.

1. The forest originally had a size of 10 acres. (An acre is a little smaller than the size of a football field without the end zones.) A year ago, the beetle killed $\frac{11}{8}$ acres of the forest. How many acres of the forest are still alive? Give your answer as an improper fraction or a mixed number.

2. To learn more about the beetle, scientists went into the remaining parts of the forest (from part 1) and placed a number of beetle traps. They decided to place at least one beetle trap every $3/4$ acres of living palm trees. What is the minimum number of traps they would need to place? (Please show your work.)

Beetle trap

3. As the beetles continued to spread, another 3.025 acres of palm trees died in this forest. What is the size of the forest now? Give your answer as a decimal.

4. Your school wants to ask the government to do more to protect these forests. Let's calculate some numbers that can help describe the situation. How many times bigger was the original forest compared to now? Give your answer as a decimal rounded to the nearest tenth.

5. A lot of people do not pay attention to environmental issues. If you could get more information about the coconut beetle, what kind of information do you think would make people pay more attention? Share your thoughts in the online comment section. ➧ (Your answer does not have to be a mathy one. =])

6. If you have seen a coconut beetle trap, please take a picture and post it online along with the location of where you saw it! ➧ Or share it with your teacher and your classmates.

Unit 3: Proportionality: Ratio and Rates

In this unit, we'll learn how to use ratios and rates to convert measurements and describe situations through avoiding lava flows and surveying invasive and non-invasive fish populations. There are five activities in this unit. *Module 6* involves representing ratios and rates by evaluating an invasive fish population. *Module 7* has two activities where students guide the keiki to safety by applying ratios and rates to a map in order to avoid Island dangers. *Module 8* explores a variety of fish populations through the use of percents. The final activity is cumulative and incorporates concepts from each of the previous activities in this unit.

Some of the activities in this unit need a ruler with centimeters to complete.

Common Core State Standards

Common Core State Standard	Module 6	Module 7	Module 8	Unit 3
6.RP.1 Understand the concept of a ratio and use ratio language to describe a ratio relationship between two quantities. *For example, "The ratio of wings to beaks in the bird house at the zoo was 2:1, because for every 2 wings there was 1 beak." "For every vote candidate A received, candidate C received nearly three votes."*	X			X
6.RP.2 Understand the concept of a unit rate a/b associated with a ratio a:b with b $\neq$ 0, and use rate language in the context of a ratio relationship. *For example, "This recipe has a ratio of 3 cups of flour to 4 cups of sugar, so there is 3/4 cup of flour for each cup of sugar." "We paid $75 for 15 hamburgers, which is a rate of $5 per hamburger."* (Note: Expectations for unit rates in this grade are limited to non-complex fractions.)	X	X		X
6.RP.3 Use ratio and rate reasoning to solve real-world and mathematical problems, e.g., by reasoning about tables of equivalent ratios, tape diagrams, double number line diagrams, or equations.	X	X	X	X
6.RP.3a Make tables of equivalent ratios relating quantities with whole-number measurements, find missing values in the tables, and plot the pairs of values on the coordinate plane. Use tables to compare ratios.		X		
6.RP.3b Solve unit rate problems including those involving unit pricing and constant speed. *For example, if it took 7 hours to mow 4 lawns, then at that rate, how many lawns could be mowed in 35 hours? At what rate were lawns being mowed?*		X		X
6.RP.3c Find a percent of a quantity as a rate per 100 (e.g., 30% of a quantity means 30/100 times the quantity); solve problems involving finding the whole, given a part and the percent.			X	X
6.RP.3d Use ratio reasoning to convert measurement units; manipulate and transform units appropriately when multiplying or dividing quantities.		X	X	X

Module 6: Representing Ratios and Rates *Activity*

When fish or other living things move to new locations, the fish can completely take over its new environment and everything that lives there. When this happens, the fish is called invasive. If the fish was there originally or if it doesn't cause harm, then the fish is called non-invasive. You are at a river in Downtown Honolulu and you are counting the number of invasive and non-invasive fish. You notice that there is a new fish in the river. Let's track the impact of the new fish on the invasive and non-invasive fish populations over the next few years.

Year	Non-invasive fish	Invasive fish
1	150	6
2	138	12
3	126	18
4	114	24
5		
6		
7		
8		
9		
10		

Examples

Non-invasive fish in Hawaii	Invasive fish in Hawaii
 'O'opu (freshwater goby)	 Bass
 'Ama'ama (striped mullet)	 Catfish
 Āholehole (spotted flagtail)	 Cichlid

1. It looks like the non-invasive fish population is dropping at a constant rate, and the invasive fish population is increasing at a constant rate. **What are the rates?** In other words, how much does the number of fish change per year? Make sure to use positive numbers to show that a number is growing and negative numbers to show that a number is shrinking.

   Non-invasive fish: __________ fish per year Invasive fish: __________ fish per year

2. Complete the previous table for the remaining six years (years 5 to 10).

3. Find the ratio of *non-invasive* to *invasive* fish on the following years. Be sure to reduce your ratios to lowest terms.

   (a) Year 3?

   (b) Year 5?

4. In what year will the ratio of *non-invasive* to *invasive* fish be $1:1$?

5. How many of each type of fish do you expect to count in year 12? Explain.

Non-invasive fish: __________ fish Invasive fish: __________ fish

6. Many plants and animals like kalo (taro), ʻulu (breadfruit), and the monarch butterfly were brought to Hawaiʻi from other places, but they are **not** considered "invasive." Why do you think that is? Can you name other plants or animals that are not from Hawaiʻi and aren't invasive either? Feel free to use the internet or work with a partner if your teacher allows.

Module 7: Applying Ratios and Rates *Activity 1*

For this activity you will need a ruler with centimeters.

A lava flow is coming! You have to help the keiki navigate this ahupua'a to get home safely.

1. Using **straight lines**, draw a path from the **keiki** to the **home**. You may use the bridge or the sandbar to get across the river, but stay out of the water and off of the lava flow. Make sure that there are no breaks in the lines that you draw.

2. Use your ruler to measure the total length of your lines in centimeters. Round your final total to the nearest centimeter.

Total length: ___________ cm

3. For this map scale, 40 feet in real life is represented by 5 centimeters on the map. How many feet in real life does each centimeter on the map represent?

4. Complete the following table.

Centimeters on the map	1	2		4	5	10		20
Feet in real life			24		40	80	88	

5. How many feet was the length of your path?

Module 7: Applying Ratios and Rates *Activity 2*

This is a follow up of the last activity. Please complete Activity 1 before trying this one! For this activity you will need a ruler with centimeters.

A lava flow is coming! You have to help the keiki navigate this ahupua'a to get home safely.

1. This time, let's work with a partner to find the **shortest** path from the keiki to the house. Using **straight lines**, draw a path from the **keiki** to the **home**. You may use the bridge or the sandbar to get across the river, but stay out of the water and off of the lava flow. Make sure that there are no breaks in the lines that you draw.

2. Find the real distance of your path. Remember, 5 centimeters of your map is 40 feet in real life. Share your solution (map and distance, in feet) and strategy with other groups!

Module 8: Percents *Activity*

Volunteers are visiting several streams in Honolulu to compare how many fish in the streams are invasive. Unfortunately, each volunteer wrote their numbers in a different formats.

Here is what they found.

Stream	Proportion of invasive fish
Kapālama	19%
Makiki	$\frac{2}{7}$
Mānoa	0.21
Nu'uanu	9%
Pālolo	$\frac{1}{5}$
Wai'alae	0.3

Examples

Non-invasive fish in Hawaii	Invasive fish in Hawaii
 'O'opu (freshwater goby)	 Bass
 'Ama'ama (striped mullet)	 Catfish
 Āholehole (spotted flagtail)	 Cichlid

1. One way to get this kind of data is to catch a lot of fish and count how many of the fish are invasive. If the volunteers caught 100 fish in each stream, how many invasive fish would you find in each stream? Round to the nearest whole number if needed. Kapālama is already done for you.

Stream	Number of invasive fish out of 100
Kapālama	19
Makiki	
Mānoa	
Nu'uanu	
Pālolo	
Wai'alae	

2. Rank the streams from the *lowest* proportion of invasive fish to non-invasive fish to *highest*.

(Lowest ratio) (Highest ratio)

3. Discuss the following questions with a partner. Write your answers and share them with others in your class.

   (a) How do you think the invasive fish got to the rivers in the first place?

   (b) How do you think some of these invasive fish end up taking over a new environment?

Unit 3: *Cumulative Activity*

The pua'a (pig) was introduced to the Hawaiian islands by the first Polynesian explorers. Although this animal is important to the Hawaiian culture, the pua'a can be pretty destructive to the environment. This is mainly because they eat a lot of food and they can eat almost anything. When they are looking for food, they may dig up and chew up native plants. After they eat, they may leave poop that contain seeds of invasive plants.

Pua'a (feral pigs)

The pua'a can often be found eating plants, roots, mushrooms, bugs, and even small mammals. Typically, the diet of a pua'a in your ahupua'a consists of 17 pounds of plants (and fungi) for every 3 pounds of animal meat.

1. Represent this typical diet as a **ratio** of plants (and fungi) to animals.

2. What **percent** of this diet would be of plants? Please explain how you got this answer.

3. What **percent** of this diet would be of animals?

4. Suppose that you found a baby pua'a in your yard. You know not to let it go because of the damage that it can do to the environment, so your family decides to keep it as a pet. If you had to prepare 500 pounds of food for this pua'a, how many pounds of plants, and how many pounds of animal meat do you need?

5. As we saw before, the pua'a in this ahupua'a eats 17 pounds of plants for every 3 pounds of animal meat. How many pounds of plants does the pua'a eat for one (1) pound of animal meat? Round your answer to the nearest tenth.

6. If a baby pua'a is eating 10 pounds of meat, how many pounds of plant should it also eat? Round to the nearest whole number.

7. The diet of a pua'a in one ahupua'a can be different from a pua'a in a different ahupua'a. Why do you think their diets might be different?

Unit 4: Equivalent Expressions

In this unit, we'll learn how to write algebraic expressions in different ways to describe situations through supporting Hawai'i's unique environment by farming. There are three activities in this unit. *Module 9* involves exploring the growth of algae through generating equivalent numerical expressions. *Module 10* focuses on generating equivalent algebraic expressions while evaluating donations to a reforestation project. The final activity is cumulative and incorporates concepts from each of the previous activities in this unit.

Common Core State Standards

Common Core State Standard	Module 9	Module 10	Unit 4
6.EE.1 Write and evaluate numerical expressions involving whole-number exponents.	X		X
6.EE.2 Write, read, and evaluate expressions in which letters stand for numbers.	X	X	X
6.EE.2a Write expressions that record operations with numbers and with letters standing for numbers. *For example, express the calculation "Subtract y from 5" as 5 – y.*			X
6.EE.2b Identify parts of an expression using mathematical terms (sum, term, product, factor, quotient, coefficient); view one or more parts of an expression as a single entity. *For example, describe the expression 2 (8 + 7) as a product of two factors; view (8 + 7) as both a single entity and a sum of two terms.*			X
6.EE.2c Evaluate expressions at specific values of their variables. Include expressions that arise from formulas used in real-world problems. Perform arithmetic operations, including those involving whole-number exponents, in the conventional order when there are no parentheses to specify a particular order (Order of Operations). *For example, use the formulas* $V = s^3$ *and* $A = 6s^2$ *to find the volume and surface area of a cube with sides of length s = 1/2.*	X	X	X
6.EE.3 Apply the properties of operations as strategies to generate equivalent expressions. *For example, apply the distributive property to the expression 3(2 + x) to produce the equivalent expression 6 + 3x; apply properties of operations to y + y + y to produce the equivalent expression 3y.*		X	
6.EE.4 Identify when two expressions are equivalent (i.e., when the two expressions name the same number regardless of which value is substituted into them). *For example, the expressions y + y + y and 3y are equivalent because they name the same number regardless of which number y stands for.*		X	X

Continued on next page

Common Core State Standard	Module 9	Module 10	Unit 4
6.EE.6 Use variables to represent numbers and write expressions when solving a real-world or mathematical problem; understand that a variable can represent an unknown number, or, depending on the purpose at hand, any number in a specified set.		X	X

Module 9: Generating Equivalent Numerical Expressions *Activity*

To help the native fish populations, you have decided to raise baby fish at home. First, you need to grow some algae for the fish to eat. You have three types of algae called Type 1, Type 2, and Type 3, and each type grows differently.

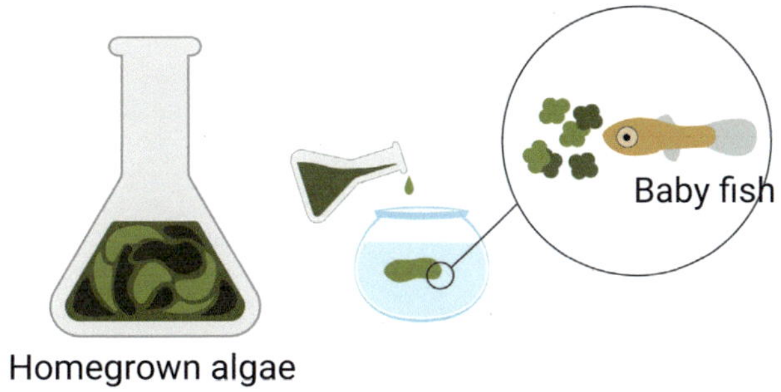

1. The Type 1 algae is tripling its population every day. After two days, its starting population has multiplied by a factor of 9 or 3^2 (in exponent form). After three days, its starting population has multiplied by a factor of 27 or 3^3 (in exponent form). Please fill out the following table.

	Number of times the Type 1 population has multiplied by	
Day	Factor	In exponent form
1	3	3^1
2	9	3^2
3	27	3^3
4		
5		
6		
7		
8		
9		
10		

2. The Type 2 algae grows more unpredictably. During the first few days, the algae quintupled (multiplied its population by a factor of 5). Then, it tripled its populations for a few days. Finally, it began to double its population every day. After a total of ten (10) days of growth, the Type 2 algae has multiplied its starting population by a factor of 12960.

   (a) Find the prime factorization of 12960 in exponential form.

   (b) Using your answer from 2a, find out how many days, out of the ten days, the Type 2 algae doubled (×2), tripled (×3), and quintupled (×5) its population.

   Double:__________ Triple:__________ Quintuple:__________

3. After D days, the Type 3 algae has multiplied by the following number.

$$1000 \div \left(1024 \div 2^D - 0.8\right)$$

   How many times did the original population of this algae multiply after ten days ($D = 10$)?

4. Rank the three types of algae from **slowest** to **fastest** growing after ten (10) days

Module 10: Generating Equivalent Algebraic Expressions *Activity*

A small high school on Hawai'i Island is raising money to buy and plant koa and 'iliahi trees.

Koa sapling

'Iliahi sapling

Below, on the left side are several **classes and a description** of how much money they have each raised. Below, on the right side are several mathematical expressions that describe the **amount of money raised**. The cost of a koa tree is k and the cost of an 'iliahi tree is h.

1. Match each class with at least one expression. A few expressions are equal so **some classes will be matched with more than one expression**.

Freshman class (grade 9):
Each student raised enough money to buy 3 koa and 2 'iliahi trees. There were 7 students. •

Sophomore class (grade 10):
The students raised enough money to buy 4 koa and 6 'iliahi trees. Then a local company decided to doubled the money they raised. •

Junior class (grade 11):
The students raised enough money to buy 21 koa trees, 14 'iliahi trees, and had \$21 left over. •

Senior class (grade 12):
The students raised enough money to buy 21 'iliahi trees and had \$14 left over. •

- $21k + 14h$
- $2(4k + 6h)$
- $7(3k + 2h)$
- $7(3k + 2h + 3)$
- $21k + 14h + 21$
- $12h + 8k$
- $21h + 14$
- $7(2 + 3h)$

2. If each koa tree cost $90, and each ʻiliahi tree cost $110, find the total amount of money that each class raised.

   (a) Freshman class

   (b) Sophomore class

   (c) Junior class

   (d) Senior class

Unit 4: *Cumulative Activity*

Many animals depend on plants, not just for food, but for shelter as well. For example, the Kamehameha butterfly is an animal that, as a caterpillar, lives on the māmaki plants of Hawai'i.

Kamehameha butterfly

Māmaki plant

There are signs that the Kamehameha butterfly is disappearing from many of its natural habitats. Let's grow a field of māmaki plants to help the butterfly thrive.

1. Right now we are building rows of māmaki plants in a 40 feet wide garden. The length of the garden depends on the number of rows. We need **eight feet for every row of māmaki**. For example, if our garden has 4 rows, then it needs a length of 32 feet.

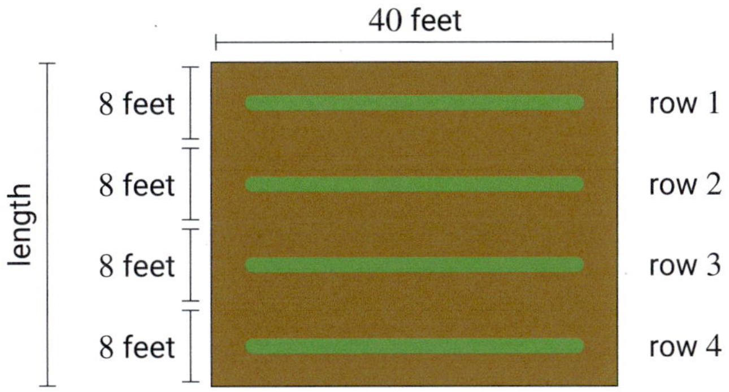

If you are building n rows, write an algebraic expression to show what the length of your garden must be.

2. Use the algebraic expression for length (in part 1) to write an expression for the area of the māmaki garden. Check with a friend to see if you have similar answers.

3. Use your expression in part 2 to determine whether the following statements are true or false.

(a) The area is the **sum** of 320 and n. True or False

(b) The area is the **product** of 40 and $8n$. True or False

(c) The area is the **quotient** of 320 and n. True or False

(d) 320 and n are **factors** in your expression in part 2. True or False

(e) 40 and $8n$ are **terms** in your expression in part 2. True or False

(f) n is a **coefficient** in your expression in part 2. True or False

4. You decide to plant 10 rows of māmaki in this garden. What is the total area of the garden? Check with a neighbor to see if you got the same answer.

5. Find the prime factorization of the answer in part 4.

6. Work with a partner to sketch, and label the widths and lengths, of at least 3 other rectangles that have the same area as the garden in part 4.

7. How can the prime factorization of part 5 help you in part 6?

8. Suppose that d number of Kamehameha butterflies land in the garden in part 4, and the Kamehameha butterflies spread out evenly in the area of the garden. Write an expression for the amount of area each of the d Kamehameha butterflies will have to themselves.

9. How much area would each Kamehameha butterfly have if there were...

   (a) 5 Kamehameha butterflies ($d = 5$)? (Round to the nearest tenth if needed.)

   (b) 9 Kamehameha butterflies? (Round to the nearest tenth if needed.)

10. There are 5 Kamehameha butterflies right now. Suppose that the butterfly population doubles every year. How many Kamehameha butterflies would there be in 4 years?

Unit 5: Equations and Inequalities

In this unit, we'll learn how to use algebraic equations and inequalities to solve problems through exploring different ahupua'a and the lengths of various Hawaiian trees in the ahupua'a. There are three activities in this unit. *Module 11* involves the use of equations and relationships to help describe the growth of native Hawaiian trees. *Module 12* focuses evaluating the distance between neighbors in a ahupua'a with the help of relationships on two variables. The final activity is cumulative and incorporates concepts from each of the previous activities in this unit.

Common Core State Standards

Common Core State Standard	Module 11	Module 12	Unit 5
6.NS.6 Understand a rational number as a point on the number line. Extend number line diagrams and coordinate axes familiar from previous grades to represent points on the line and in the plane with negative number coordinates.		X	X
6.NS.6b Understand signs of numbers in ordered pairs as indicating locations in quadrants of the coordinate plane; recognize that when two ordered pairs differ only by signs, the locations of the points are related by reflections across one or both axes.		X	
6.NS.6c Find and position integers and other rational numbers on a horizontal or vertical number line diagram; find and position pairs of integers and other rational numbers on a coordinate plane.		X	X
6.NS.8 Solve real-world and mathematical problems by graphing points in all four quadrants of the coordinate plane. Include use of coordinates and absolute value to find distances between points with the same first coordinate or the same second coordinate.		X	
6.EE.5 Understand solving an equation or inequality as a process of answering a question: which values from a specified set, if any, make the equation or inequality true? Use substitution to determine whether a given number in a specified set makes an equation or inequality true.	X		X
6.EE.6 Use variables to represent numbers and write expressions when solving a real-world or mathematical problem; understand that a variable can represent an unknown number, or, depending on the purpose at hand, any number in a specified set.	X		X
6.EE.7 Solve real-world and mathematical problems by writing and solving equations of the form $x + p = q$ and $px = q$ for cases in which p, q and x are all nonnegative rational numbers.		X	X

Continued on next page

Common Core State Standard	Module 11	Module 12	Unit 5
6.EE.8 Write an inequality of the form $x > c$ or $x < c$ to represent a constraint or condition in a real-world or mathematical problem. Recognize that inequalities of the form $x > c$ or $x < c$ have infinitely many solutions; represent solutions of such inequalities on number line diagrams.	X		
6.EE.9 Use variables to represent two quantities in a real-world problem that change in relationship to one another; write an equation to express one quantity, thought of as the dependent variable, in terms of the other quantity, thought of as the independent variable. Analyze the relationship between the dependent and independent variables using graphs and tables, and relate these to the equation. *For example, in a problem involving motion at constant speed, list and graph ordered pairs of distances and times, and write the equation $d = 65t$ to represent the relationship between distance and time.*	X	X	X

Module 11: Equations and Relationships *Activity*

This house is built right next to a forest. Near the house are some fast-growing trees.

After x years, the height of the trees (in feet) are given by the following expressions.

Tree	Height after x years
Mai'a (banana)	$2x + 2$
Loulu (palm)	$\frac{3}{2}x$
Niu (coconut)	$3x + 1$
'Ohe (bamboo)	$\frac{5}{2}x + 3$

1. When writing equations and inequalities about these trees, we can replace "the height of the tree" with the expression that describes it.
   For example,
   "the height of the 'ohe is greater than or equal to 15 feet" can be written as
   $$\frac{5}{2}x + 3 \geq 15$$
   and
   "the height of the 'ohe is the height of the niu" can be written as
   $$\frac{5}{2}x + 3 = 3x + 1.$$
   Write the equations and inequalities for each of the following sentences. You **do not** need to solve them.

   (a) The height of the loulu is 3 feet.

   (b) The height of the niu is half of the height of the mai'a.

   (c) The height of the mai'a is less than the height of the 'ohe.

   (d) The height of the niu is 2 feet more than the height of the 'ohe.

2. How tall will each tree be after 2, 4, and 6 years?

Number of years	Tree height (feet)			
	Mai'a	Loulu	Niu	'Ohe
x	$2x+2$	$\frac{3}{2}x$	$3x+1$	$\frac{5}{2}x+3$
2		3		8
4			13	
6	14			

3. Draw the trees after 6 years ($x = 6$).

4. There is a 15-foot tall power line connected to this house. We need to make sure that the trees doesn't grow as tall as the power line.

   (a) Which tree(s) would reach the power line in 6 years?

   (b) What are three things that we can do to the trees if they get too tall?

Module 12: Relationships in Two Variables *Activity*

You and your ‘ohana (family) live in this ahupua‘a. Here is a map of your ahupua‘a.

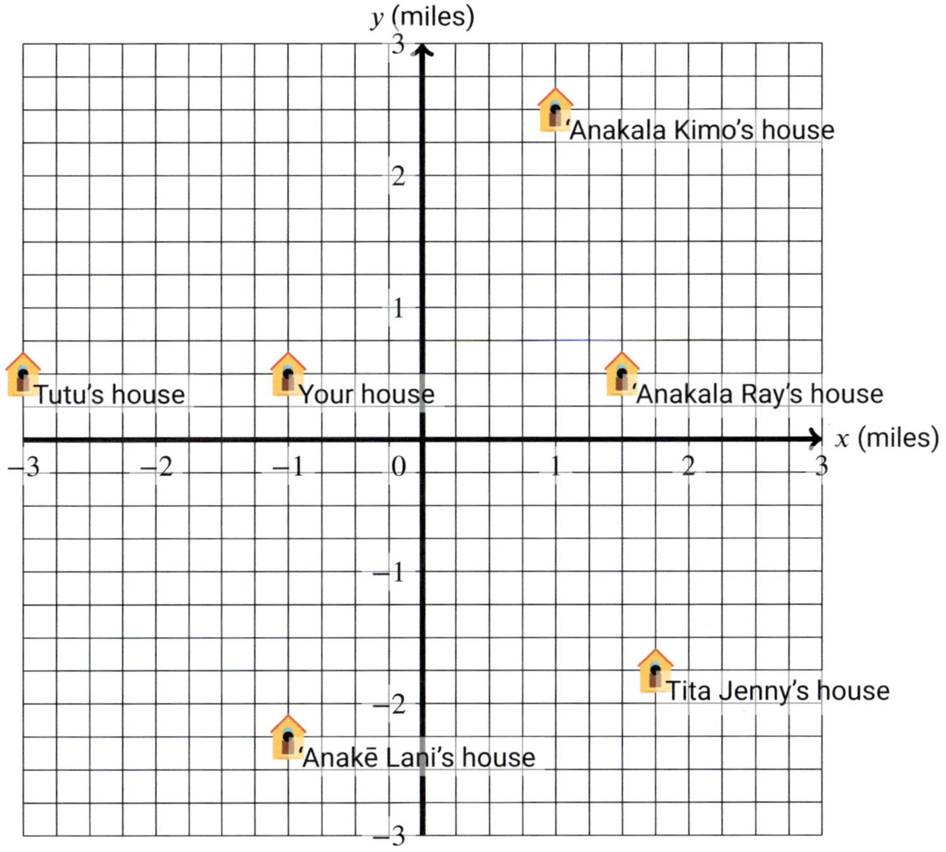

1. If you only move left, right, up, or down (not diagonal or in a curve), how many miles away **from your house** does each relative live?

   (a) Tutu

   (b) ‘Anakala Ray

   (c) ‘Anakē Lani

   (d) ‘Anakala Kimo

   (e) Tita Jenny

2. Suppose that you can jog 4 miles per hour. Write an equation describing the distance you jog (d) after t hours.

3. Using your equation from part 2 and the distances from part 1, calculate how many hours, t, it takes to jog from your house to...
   (a) Tutu's house
   (b) ʻAnakala Ray's house
   (c) ʻAnakē Lani's house
   (d) ʻAnakala Kimo's house
   (e) Tita Jenny's house

Unit 5: *Cumulative Activity*

A pond in your ahupua'a is badly polluted. You want to clean it up but, it is big pond and you wont be able to clean it alone so you decide to ask some of your neighbors to help out.

1. If x is the number of neighbors who will help, write an expression for the total number of workers (including yourself) who will clean up the pond.

2. The percent of pollution remaining in the pond can be given by a simple algebraic equation: "The *percent of pollution* is 100% divided by the *total number of workers* cleaning it up." Write this algebraic equation for percent of pollution, y in terms of number of neighbors, x.

3. Complete the following table (round to one decimal place, if needed).

Number of neighbors, x	Percent polluted, y
0	100
1	
2	
3	
4	20
5	
6	
7	
8	11.1
9	
10	

4. Graph the data from the previous table.

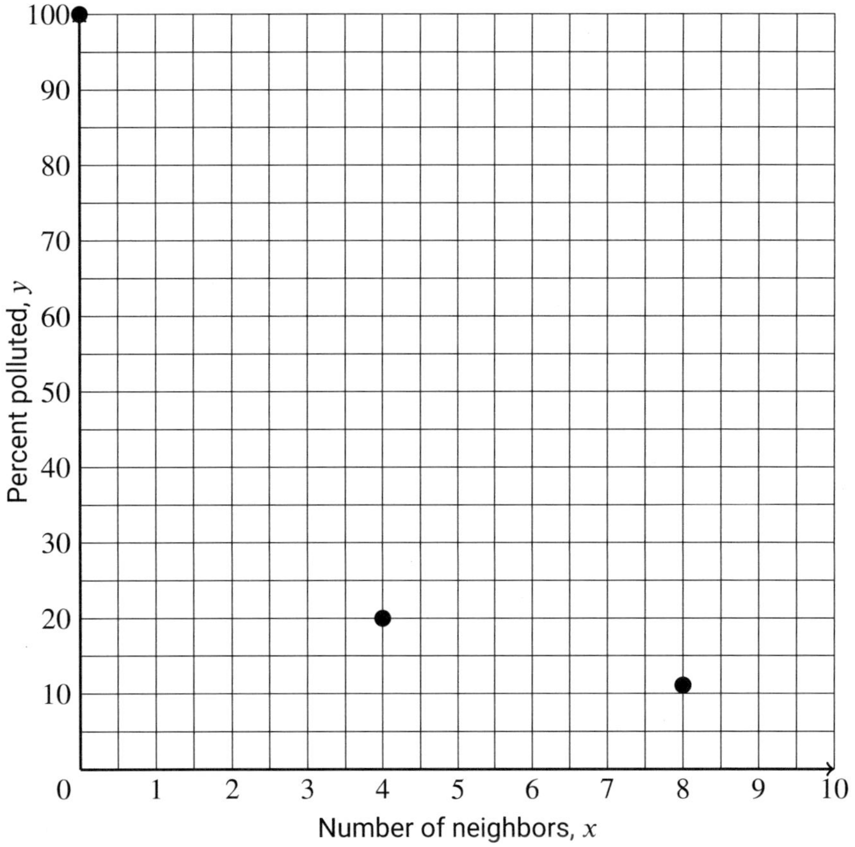

5. The graph in part 4 only shows Quadrant I. Do you think there is important information in any of the other three quadrants? Why or why not? Feel free to discuss with your friends.

Due to the pollution, there hasn't been a lot of animals visiting this pond.

6. Birds will return to the pond if it is less than 15% polluted. Modify the equation from part 2 to describe this as an inequality in terms of x.

7. Use the previous table or graph to answer this question. How many neighbors need to help in order to get this pond clean enough for birds to return?

At the end of the cleanup effort, the birds started to return to the pond. Here are the number of birds, b, d days after the cleanup.

Days (d)	0	1	2	3	4	5	6
Birds (b)	0	3	6	9	12	15	18

8. Which variable, b or d, is the independent variable, and which variable is the dependent variable? How do you know?

9. Write the equation for b in terms of d.

10. With a partner or in the online comment section, describe a place that you have been to that was really polluted. Describe what it would look like if people took the time to clean it up.

Unit 6: Relationships in Geometry

In this unit, we'll learn how to draw shapes on a grid and calculate length, area, and surface area through surveying Hawaiian land, exploring the lengthy process of kapa making, and building a hale. There are four activities in this unit. *Module 13* involves helping a land surveyor with the use of area and polygons. *Module 14* focuses on distance and area in the coordinate plane to help wahine make kapa. *Module 15* explores how to build a hale with the help of surface area and volume of solids. The final activity is cumulative and incorporates concepts from each of the previous activities in this unit.

For some of the activities in this unit, a yard stick or measuring tape would be helpful but is not required.

Common Core State Standards

Common Core State Standard	Module 13	Module 14	Module 15	Unit 6
6.NS.6b Understand signs of numbers in ordered pairs as indicating locations in quadrants of the coordinate plane; recognize that when two ordered pairs differ only by signs, the locations of the points are related by reflections across one or both axes.		X		X
6.NS.8 Solve real-world and mathematical problems by graphing points in all four quadrants of the coordinate plane. Include use of coordinates and absolute value to find distances between points with the same first coordinate or the same second coordinate.		X		
6.EE.7 Solve real-world and mathematical problems by writing and solving equations of the form $x + p = q$ and $px = q$ for cases in which p, q and x are all nonnegative rational numbers.	X			
6.G.1 Find area of right triangles, other triangles, special quadrilaterals, and polygons by composing into rectangles or decomposing into triangles and other shapes; apply these techniques in the context of solving real-world and mathematical problems.	X	X	X	X
6.G.2 Find the volume of a right rectangular prism with fractional edge lengths by packing it with unit cubes of the appropriate unit fraction edge lengths, and show that the volume is the same as would be found by multiplying the edge lengths of the prism. Apply the formulas $V = l\ w\ h$ and $V = b\ h$ to find volumes of right rectangular prisms with fractional edge lengths in the context of solving real-world and mathematical problems.			X	
6.G.3 Draw polygons in the coordinate plane given coordinates for the vertices; use coordinates to find the length of a side joining points with the same first coordinate or the same second coordinate. Apply these techniques in the context of solving real-world and mathematical problems.		X		X

Continued on next page

Common Core State Standard	Module 13	Module 14	Module 15	Unit 6
6.G.4 Represent three-dimensional figures using nets made up of rectangles and triangles, and use the nets to find the surface area of these figures. Apply these techniques in the context of solving real-world and mathematical problems.			X	X

Module 13: Area and Polygons *Activity*

A land surveyor is a person who measures and maps out the details of a piece of land. Let's survey two pieces of land.

The first piece of land is from the ahupua'a of Wai'anae. We made it easier to measure the land by breaking it up into right triangles and rectangles. Some of the sides and areas (the numbers in a circle) have been measured.

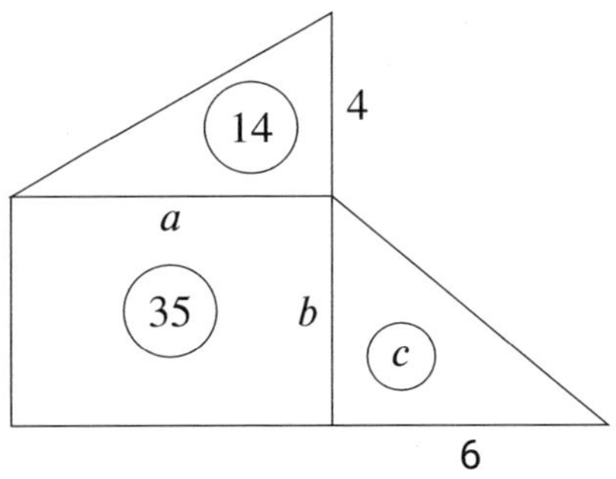

1. Find the two missing sides and then the missing area.

   (a) Side a:

   (b) Side b:

   (c) Area c:

2. Find the total area of the land in Wai'anae.

Here is a piece of land from the ahupua'a of Waiau.

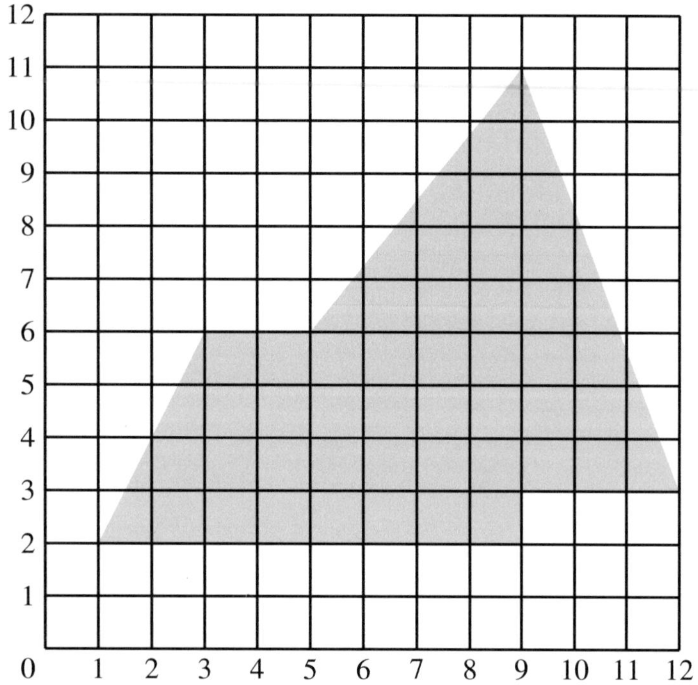

3. Break up the Waiau land into right triangles and rectangles by drawing on the figure above.
4. Find the total area of the land in Waiau.

5. Which piece has a larger total area, Wai'anae or Waiau?

Module 14: Distance and Area in the Coordinate Plane *Activity*

Kapa is a cloth made from the wauke (paper mulberry) plant. It takes a lot of hard work to make kapa. In old Hawai'i, Hawaiians gathered strips of young wauke bark and repeatedly battered it until the fabric was soft. Then they had to forage for more pieces, line them up next to each other, and pound the edges until the different strips connected. It would take a very long time to make a large piece that was soft, strong, flat, and had the same thickness throughout. After the pieces were beaten together, they had to be dried in the sun before being made into clothing, furniture, or wraps.

The making of kapa was very important to the ancient Hawaiians. It was said that the demi-god, Maui, threw a hook into the Sun to slow it down so that his mom, Hina, could dry her kapa that she had worked so hard to make.

A wahine in our village is making kapa and she needs a place to dry it in the sun. However, we need to build a fence to keep animals and kolohe kids away from the kapa while it is drying.

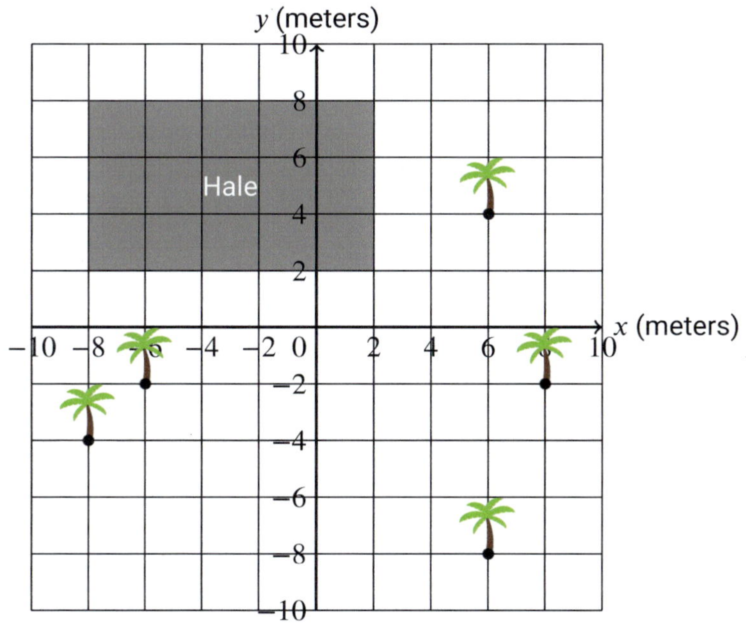

1. Draw a triangle connecting the three points on the map above: $(2,2)$, $(8,-2)$, and $(6,-8)$.
2. Let's draw a rectangle around the triangle in part 1 by connecting the following points: $(2,2)$, $(8,2)$, $(8,-8)$, and $(2,-8)$. You should now see four triangles: the original one and three right triangles.
3. Use the rectangle and the three right triangles in part 2 to find the area of the original triangle in part 1. Show your work below.

4. The wahine needs at least 40 square meters to dry all of the kapa. Is there enough area in the triangle?

5. Draw a new drying area by connecting the following points in order.
$(2,2) \to (-2,1) \to (-4,-2) \to (-6,-2) \to (-6,-4) \to (-8,-4) \to (-8,-8) \to (2,-8) \to (2,2)$

6. Find the area of the polygon that you drew in part 5.

7. Is there enough area in this polygon for the wahine to dry her kapa?

Module 15: Surface Area and Volume of Solids *Activity*

A yard stick or measuring tape could be helpful in this activity.

In parts of Hawai'i that stay warm all year long, the ancient Hawaiians rarely spent much time in their hale. The doors on hale were often made very small to keep pets and pests out, and the insides of the hale were often very warm and dark. The hale was mostly used for storage and rituals, except for cold and rainy weather or special occasions.

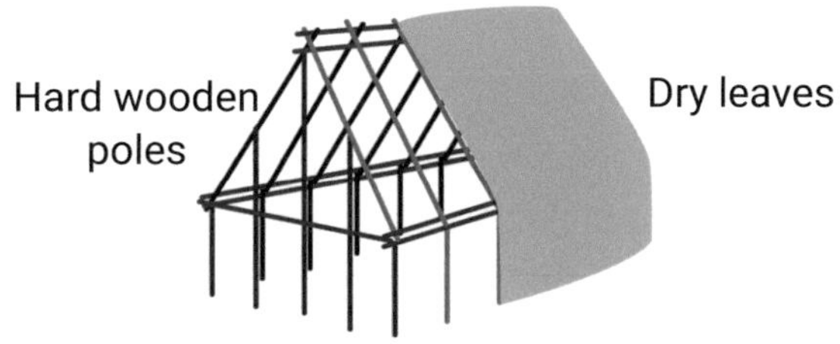

A hale under construction

We are building a hale. Let's first look at it without a roof.

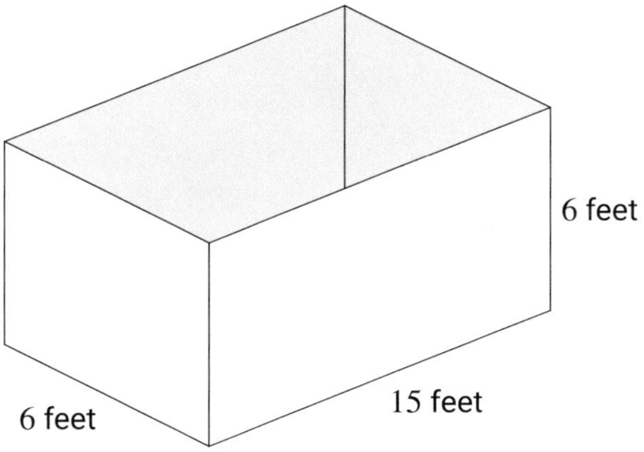

1. Find the volume of the rectangular prism.

2. How many of your classmates do you think can fit in a hale of this size? Discuss with a friend. Walking around the classroom with a yard stick or measuring tape while imagining a space this size can also help.

3. Divide your answer in part 1 by your answer in part 2 to find out how much space each of your classmates would take up in this hale. Round to the nearest tenth.

Now let's look at the hale with its roof.

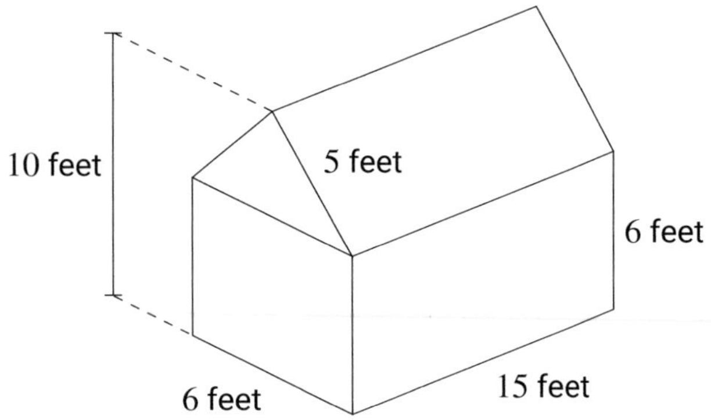

4. Sketch the four walls and the four sides of the roof as a geometric net. Label each edge with its length. You only need to use triangles, squares, and rectangles.

5. We need to cover the four walls and the four sides of the roof with thick leaves. Find the total surface area of these eight pieces.

6. When a hale needs to be built, many people must come together to build it. When was the last time you built something with your hands? What did you make?

Unit 6: *Cumulative Activity*

Ivy gourd is an invasive vine that covers and kills other plants. A bird pooped in a field next to your ʻili (neighborhood) and three ivy gourd plants are starting to grow from the seeds in the bird poop.

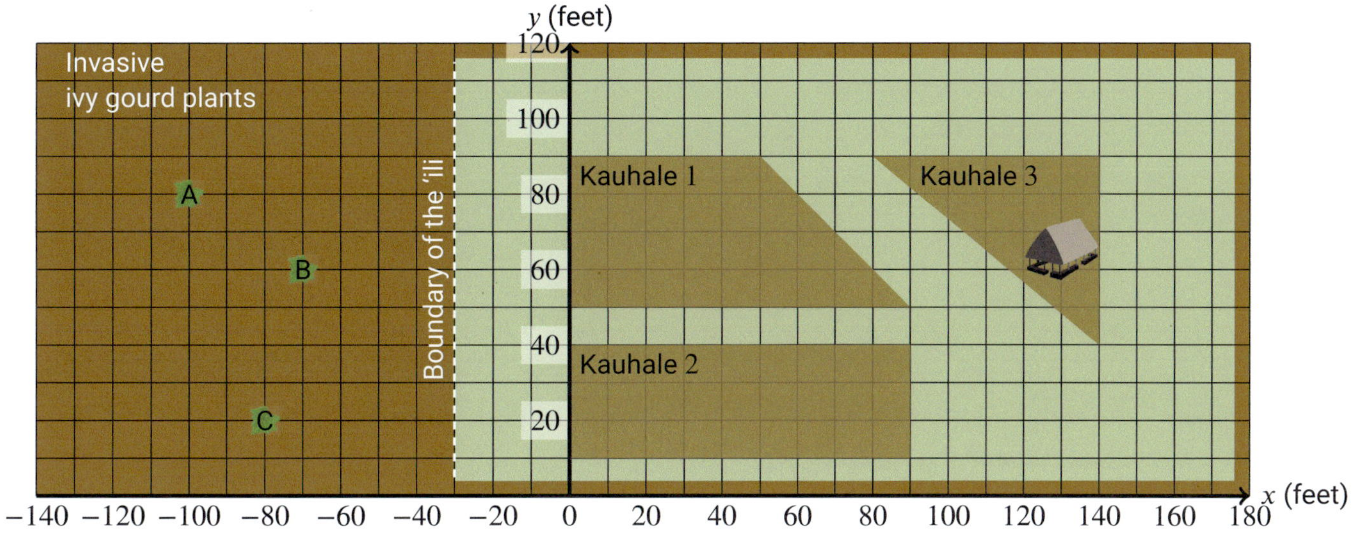

1. How far away is each plant from the boundary of the ʻili?
   (a) Plant A:
   (b) Plant B:
   (c) Plant C:

2. Look at the coordinates of each ivy gourd plant. If we changed the signs of each x-coordinate to it's opposite, would the plant be moved into a kauhale? If so, which kauhale would it be in?
   (a) Plant A:
   (b) Plant B:
   (c) Plant C:

3. After the ivy gourd plants enter your ʻili, they quickly cover the three kauhale, where the hale (houses) are built. What is the **total** area of the three kauhale?

4. In kauhale 3, the ivy gourd is invading a hale. First, it covers the floor of the hale, then the walls and the roof. What is the surface area of the covered hale?

Hale

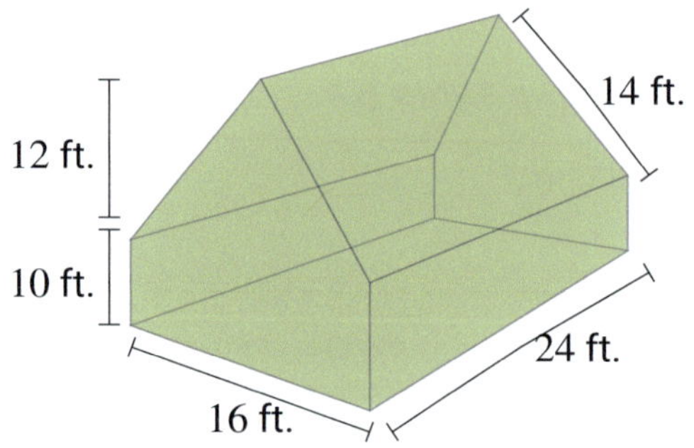

Hint: The front and back walls is a triangle on top of a rectangle. The roofs, side walls, and floor are rectangles.

5. Most plants are good for the environment and good for people, but certain plants, when grown in wrong places, can cause a lot of trouble. Describe a plant in your neighborhood or school that you think is out of place. Describe why you think it might be bad for your neighborhood. You may also find or draw a picture of it. Please share with your partner or in the online comment section. (If sharing online, please don't post pictures of people's faces or anything that can be used to find a personal address.)

Unit 7: Measurement and Data

In this unit, we'll learn how to use statistics to summarize and display data through evaluating the shape and length of a Hawaiian spear and the exploration of the Big Island. There are two activities in this unit. *Module 16* explores the Hawaiian spear, pololū, through displaying, analyzing, and summarizing data. The final activity is cumulative and incorporates concepts from each of the previous activities in this unit.

Common Core State Standards

Common Core State Standard	Module 16	Unit 7
6.SP.1 Recognize a statistical question as one that anticipates variability in the data related to the question and accounts for it in the answers. *For example, "How old am I?" is not a statistical question, but "How old are the students in my school?" is a statistical question because one anticipates variability in students' ages.*	X	X
6.SP.2 Understand that a set of data collected to answer a statistical question has a distribution which can be described by its center, spread, and overall shape.	X	X
6.SP.3 Recognize that a measure of center for a numerical data set summarizes all of its values using a single number, while a measure of variation describes how its values vary using a single number.	X	X
6.SP.4 Display numerical data in plots on a number line, including dot plots, histograms, and box plots.	X	X
6.SP.5 Summarize numerical data sets in relation to their context, such as by:	X	X
6.SP.5a Reporting the number of observations.	X	X
6.SP.5b Describing the nature of the attribute under investigation, including how it was measured and its units of measurement.	X	X
6.SP.5c Giving quantitative measures of center (median and/or mean) and variability (interquartile range and/or mean absolute deviation), as well as describing any overall pattern and any striking deviations from the overall pattern with reference to the context in which the data were gathered.	X	X
6.SP.5d Relating the choice of measures of center and variability to the shape of the data distribution and the context in which the data were gathered.	X	X

Module 16: Displaying, Analyzing, and Summarizing Data *Activity*

The ancient Hawaiians were very fortunate to have plenty of woody plants around them. The wood from koa trees and kauila shrubs, in particular, are very solid. This quality made them very important for making tools and weapons in ancient Hawai'i. Let's use statistics to figure out the typical length of a Hawaiian spear called the pololū.

Pololū

There is a collection of ancient pololū in a museum. Let's take a look at their lengths (in feet).

11	15	8	7	11	10
9	6	12	10	13	10
9	9	9	8	7	12

1. Plot the lengths on a dot plot.

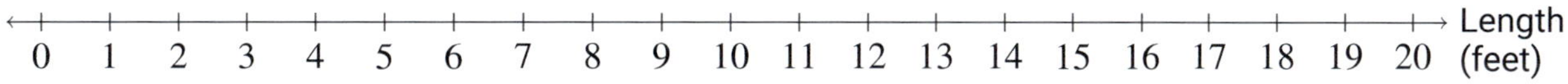

2. Describe the shape of this data distribution.

3. Which pololū length was the most common? How many of the ancient pololū were this length?

4. Find the mean and median of this data. Round to the nearest tenth.

5. Could any of your answers from parts 3 and 4 be used to represent the typical length of a pololū? Why or why not? If not, which one of those answers do you think is best?

6. How do you think life would have been different in ancient Hawai'i if woody plants were rare? Please share with your partner or in the online comment section.

Unit 7: *Cumulative Activity*

King Kamehameha the Great was born in the moku of Kohala, on the northern end of Hawai'i Island. Today, Kohala is split into North Kohala and South Kohala. Let's look at the sizes of some ahupua'a in North Kohala.

Ahupua'a	Area (km^2)
'A'amakāō	7.06
'Āinakea	2.18
'Āpuakaohau	1.19
'Āwini	8.00
Hala'ula	2.45
Hālawa	6.42
Halelua	3.03
Hana'ula	2.53
Hāwī	1.29
Honokāne	19.03
Honomaka'u	1.46
Honopueo	3.30

1. Make a box and whisker graph using the area of each ahupua'a.

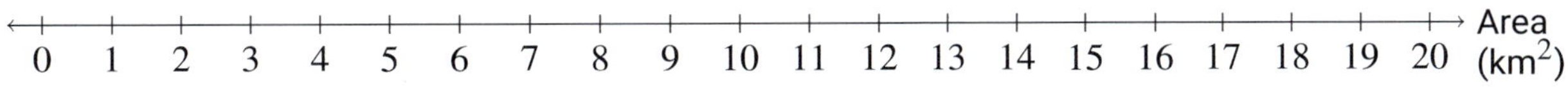

2. Calculate the range and interquartile range of the areas.

3. Calculate the mean of the areas. Round to two decimal places.

4. If we removed Honokāne from the table, how would this affect your box and whisker graph?

5. Would removing Honokāne from your table affect the mean? If so, how would it be affected? (You do not need to calculate the mean again.)

6. Make a histogram using the area of each ahupuaʻa.

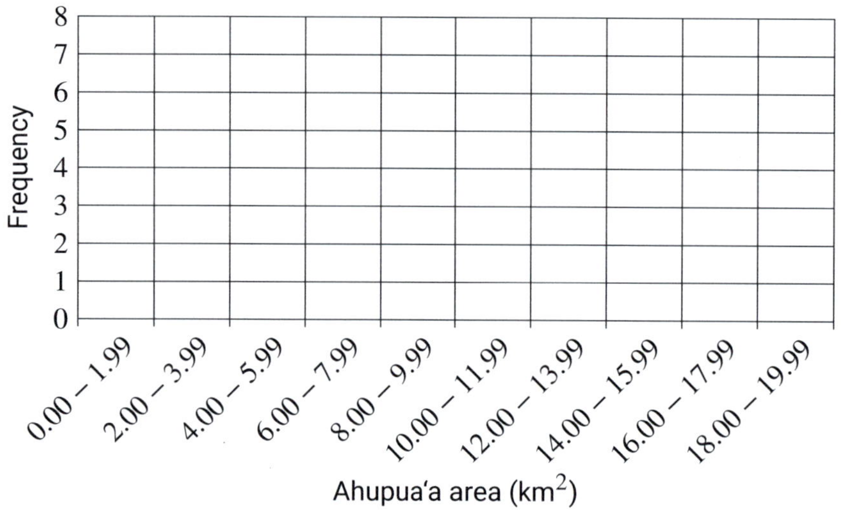

7. Which measure of center better describes the typical size of an ahupuaʻa in North Kohala–the median or the mean? Explain.

8. Choose an ahupuaʻa that you are familiar with but do NOT live in. Share something that you really like about that ahupuaʻa. Please share with your partner or in the online comment section.